Authentic Recipes
From the Golden Age of
The Soda Fountain
Dedicated to the Memory of
The Blackberry Bloom, Catawba Cobbler,
And Siberian Flip—
And to the Heroes Who Made Them

HOWARD JOHNSON'S PRESENTS OLD TIME ICE CREAM SODA FOUNTAIN RECIPES OR HOW TO MAKE A SODA FOUNTAIN PAY

1971
Winter House Ltd
New York

Manufactured in the United States of America
ISBN
0-87806-020-0

The 1903 edition of
How to Make a Soda Fountain Pay
was rescued from obscurity by David Hoffman.
For help in revising and adapting the
text, the editors of Winter House
gratefully acknowledge
Charlotte Adams, Leslie Elliott,
Diane Harris, Naomi Komisar,
and the staff
of Howard Johnson's.

CONTENTS

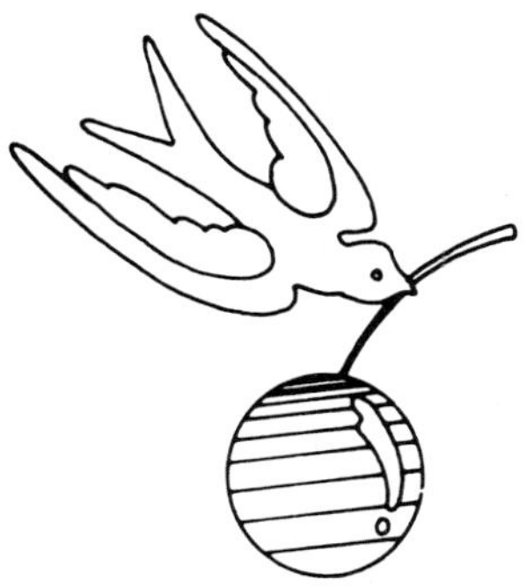

"Eternal vigilance is the price of a good soda.
Have an eagle eye, therefore.
Never pass the fountain without noting whether
the glasses are clean,
the fountain and slab immaculate,
the holders properly polished, the lemons fresh,
the eggs clean, large, and brown—
in fact, that everything is in proper shape
to tempt a thirsty mortal or to make a mortal
thirsty."

The ice cream chair

THE GREATEST SODA FOUNTAIN HERO OF THEM ALL

I scream
You scream
We all scream for
Ice cream!

Hot summer days on hot summer streets all over America, and down at the corner kids in knickers and starched white collars are lining up at the soda fountain. They slide their nickels across the gleaming counter surface and back comes—you name it: A Catawba Cobbler? A Cleveland Gem? How about the house special—the country's finest chocolate ice cream soda!

With the requested item comes the grin of America's summertime hero: the soda jerk. He tended his fountain—that chromed-and-mirrored source of all good things—with meticulous care; for he belonged to a nationwide corps of craftsmen who dreamed up new concoctions, amended old ones, and who for over fifty years had as their handbook the volume on which this one is based: *The American Dispensers Book...Containing Choice Formulas for Making Soda Water Syrups and Fancy Drinks, or: How to Make a Soda Fountain Pay.*

First published in 1863, it was issued sporadically through the years until it contained a goldmine of prize recipes, all contributed by soda fountain experts the country over. There were recipes for ice cream and recipes for sherbet; recipes for syrups with exotic and wonderful flavors—like violet and rose and orgeat; there were recipes for sodas and punches and for strange new inventions like milk shakes and something originally called a "Sunday." (That name raised church-goer eyebrows in the Midwest town where it was invented; it was changed to "Sundae.") Other concoctions were christened with romantic names—like "Jacqueminot" after a rose grown in turn-of-the-century gardens. Others had ingredients even more splendid than their names. There was the "Buffalo," invented perhaps at a shiny soda fountain in that very town. It contained syrup and soda of course, and a dash of vanilla and a splash of grape juice and a scoop of sherbet and a chunk of pineapple and a slice of orange and a maraschino cherry and a strawberry and a sprig of mint! There were egg creams with real eggs, and Coca-Cola hadn't been trademarked yet. Everyone made their own, spelling it differently every time and adding a dash of celery extract perhaps, or a shake of ginger.

Everybody came to the ice cream parlors. Men donned moustache protectors to keep their "handlebars" tidy and leaned over the counter to savor "Square Meals." The kids saved their pennies for Saturday ice cream cones, and teenagers held hands around "White Mountain Foam" for two. Ladies sat on ice cream chairs at corner tables and sipped "Charlotte Russes" through straws. And sometimes women came in alone; for years the ice cream parlor was one public place where an unescorted lady could be served without endangering her reputation.

Ice cream parlors were going up in every town and on every street corner in every large city. And behind many a counter there was a copy of *How to Make a Soda Fountain Pay.* The novice soda jerk flipped through its pages to discover not only how to make his staple, ice cream, but how to manage his business. For the little dispenser book contained a list of operating hints that reads like a guide to the old American Way of Life. Be clean. Be neat. Use only the freshest ingredients. Serve only the finest products. You will be rewarded with a successful—and profitable—soda fountain.

In 1925, one more young man joined the ranks of soda fountain entrepreneurs. He had fought in France during the Great War, returned home to buzz New England highways in a yellow

Stutz Bearcat selling cigars for his father's firm, and now he was in business for himself. "Howard D. Johnson," read the sign over the Wollaston, Massachusetts, store front. "Newsstand, lending library, homemade ice cream." He served sodas and sundaes and flips and fizzes—and cones topped with the best ice cream in town. He knew because he made the ice cream himself. But the parlor seated only ten people and so, when summer came, Johnson packed some drums of ice cream off to the Wollaston beach and set up a stand.

"We took the drums down there in taxicabs," he remembers. "We kept running out and I'd send back to town for more and call out to the crowd, 'Stand by, everybody. More ice cream on its way.' " They did stand by; they stood by in crowds so large that twelve policemen were called out from town to keep order. And then, one blazing August day on the Wollaston beach, Howard Johnson sold 14,000 ice cream cones. The greatest soda fountain hero of them all had arrived. Here, clearly, was a man who had learned how to make a soda fountain pay.

"When I bought the shop in Wollaston, the soda fountain was just a side line. The ice cream had always come from a local manufacturer, and every time we opened up a drum of vanilla in front of a customer, the fumes gave away the artificial flavoring. I could see nobody was going to come back for my ice cream unless it tasted like good food—tasted at least as good as the homemade ice cream they all remembered." But he knew what good ice cream tasted like too. His mother had a herd of twenty Jersey cows when he was growing up, and on Sundays she made ice cream using that rich Jersey cream and fresh garden strawberries. *That* was ice cream! But he remembered something else—remembered when the cigar-makers who produced the "Johnson's Private Stock" cigar especially for his father had cut the quality of the tobacco. People had shrugged, and stopped buying. If he were to put his name on ice cream, it had to be the best ice cream he could get. And that meant he would have to make it himself. So, every evening after the shop was closed, he had an expert come up from Boston to teach him the trade.

"Finally I thought I had it all down and made up my very

first run. I'll never forget the next morning. Served my first customer and then casually wandered down to him and said, 'Say, what did you think of that ice cream?' 'Be just great,' he said. 'If you could get the sand out of it.' It was gritty! And I started again. Before long, people crowded that shop. They formed lines down the block. Business was so good it was getting impossible to keep the place in any sort of shape. Hot summer days I'd see those kids trailing ice cream all over the place—and we'd be so busy that the back of the fountain was pretty bad too. I used to raise my arms and shout, 'OK, everybody. We're closing shop for ten minutes while we clean up.' They'd all move out to the sidewalk, and in ten minutes flat we'd have that place clean as a whistle."

It *was* good ice cream. It was the very best ice cream he could possibly make. He was selling it all—and then stretching the working day even further to cover the needs of that first "branch" on the Wollaston beach. If Johnson wasn't on his way to the basement to fill the ice cream freezers, ten in a row, he was on his way up to the soda fountain, one ear cocked to catch the slowing tempo of a freezer motor that would announce another batch nearly ready to go. Twelve hours, fourteen, eighteen. Twenty-one. Finally he took a trusted employee down to the basement with him, handed him an alphabetized notebook, and swore him to secrecy. "Mr. Johnsons" was written across the cover in a clear, round hand. "Personal." Inside were the perfected formulas for twenty-eight flavors of ice cream. Howard Johnson went back upstairs and began to look beyond the confines of Wollaston.

The little shop is gone now, and so is the stand down by the beach. They have been replaced by more than 900 restaurants, over 400 motor lodges, and thousands of employees who make ice cream and other food products. Behind them is a world-famous company headed by Howard B. Johnson, the son of the founder.

"I never saw a copy of the dispenser book," his father says. "Wish I had. I had to figure all this out for myself. The idea for a house special, for instance. Chocolate ice cream was my favorite—still is—and so we served a chocolate ice cream soda that people thought was terrific. I started with my own chocolate syrup, added fresh cream, and mixed it with a fine stream of soda. Then I filled it up with soda and on the very top I put a scoop of chocolate ice cream. I put that scoop on the very edge of the glass so it seemed to overflow the drink. We sold so many that after a while I had the glasses marked for all my restaurants so every customer would know he was getting just the right amount of syrup and just the right amount of cream. No matter where he was—Quincy, Route 40, Chicago."

And cleanliness, neatness, attentive service? Absolutely. For years he haunted his restaurants checking the chocolate ice cream sodas for cream, checking the door jambs for smudges, the parking lots for stray papers. Value was always a selling point. "Ice cream cones were already in existence when I came along, of course. They used to make these flat, waffle-like things and curl them up with a hot iron for the cone shape. Well, I would order cones so big for the Wollaston shop that the kids had to use two hands to hold them. Somebody once looked at that ice cream cone and figured it was costing me eight and a half cents—and I sold it for a dime. Not what you'd call a great profit," he laughs, 'but it made me and that shop and that ice cream famous."

The chapters that follow are filled with items from the book whose formulas, recipes and techniques Howard Johnson learned on his own. There are slurpy sodas, fizzes, phosphates, egg drinks—and some of those fabled super whammies that graced the turn-of-the-century soda fountain. And, those of you still inspired by the profit motive can look among these pages for bits of wisdom extracted from the gone-but-not-forgotten *How to Make a Soda Fountain Pay.*

"It is at times useful to know that—
a wineglass holds 2 ounces,
a tablespoon, ½ ounce,
a dessert spoon, ¼ ounce,
a teaspoon, ⅛ ounce or dram.
Also that a teacupful of sugar weighs ½ lb.
And 3 tablespoonfuls, ¼ lb.
16 fluid ounces equals 1 pint,
2 pints equals 1 quart,
4 quarts equals 1 gallon."

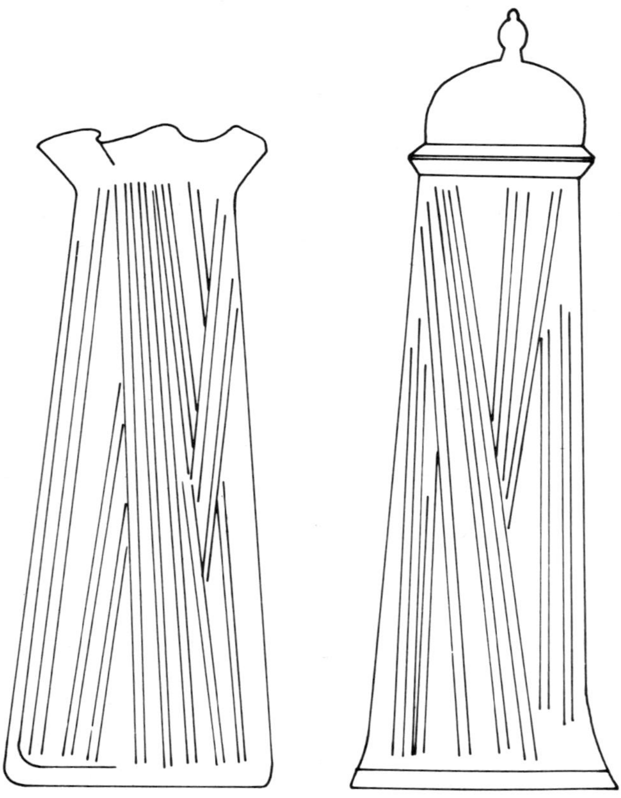

Straw vases

STARTING WITH SYRUPS

Syrups are the basis of a soda fountain, both yours and the turn of the century's. For, very simply, once you have a syrup, you also have a drink. Pour an ounce or two of syrup, fill the glass half full of soda, stir; fill to the top with soda and—presto! New Drink. (For home use, bottled club soda is the present-day substitute for the "soda" or "carbonated water" of yore; the tiny amount of salts in club soda can be ignored as you follow the recipes given.) Add ice cream and again—presto! Ice Cream Soda.

A tour through any supermarket or specialty shop will pile your shopping cart with a fine assortment of syrups: chocolate, vanilla, strawberry; pineapple syrup, lemon, raspberry, blackberry, claret—these and more, all readily available, are the basic syrups. They saw the soda jerk through the week, and will see you through most recipes. But he also learned to mix, to mingle, to add a dash of something special. And you can too.

So, stop off at a spice and seasoning rack for extracts: almond, cinnamon, peppermint—there are nearly a hundred. At a wine and liquor shop, ask for Rhine wine, Madeira, crême de menthe ...and once you're home, follow this recommendation: Combine. This, for example, is a famous recipe that found its way into the pages of *How to Make a Soda Fountain Pay:*

DON'T CARE SYRUP

"Don't Care" is a fancy name, and any syrup will do. We sometimes advise a mixture of the most popular syrups, but the following formula is said to be very fine:

Pineapple Syrup	1 pint
Strawberry Syrup	1 pint
Extract of Vanilla	1½ ounces
Port Wine	½ pint
Simple Syrup	1 gallon

Following the rule-of-thumb that a gallon of syrup, 128 ounces, will make as many as 128 drinks, you now have on hand enough syrup for six months—or one enormous party. But there were simple proportional combinations too, all of which the home soda jerk can concoct by blending syrups from supermarket shelves.

AMBROSIA SYRUP

Equal parts raspberry and vanilla syrups with a splash to taste of Rhine wine.

ALMOND CHOCOLATE SYRUP

To chocolate syrup add extract of almond to taste—a drop or two will do.

NECTAR SYRUP

Equal parts vanilla, orange, and strawberry syrups—but a dash of almond extract and a splash of Madeira wine may also be added. Equal parts nectar and cherry syrup give you—Cherry Nectar syrup, of course.

PEACH AND PINEAPPLE OR PEACH AND ORANGE SYRUP

Double the volume of either orange or pineapple syrup with the meat of crushed, skinned peaches. Delicious!

PISTACHIO SYRUP

To vanilla syrup add a dash to taste of almond extract and color with green food coloring.

SHERBET SYRUP

Equal parts orange, pineapple, and vanilla syrups.

VANILLA COFFEE SYRUP

To two parts coffee syrup add one part vanilla syrup.

VIENNA CHOCOLATE SYRUP

To chocolate syrup add extract of cinnamon to taste.

But for those brave souls who would like to try it all from scratch, here—from *How to Make a Soda Fountain Pay*—are formulas for making syrups.

The nearly nation-wide practice at the turn of the century was to keep in ready supply gallon jugs of "simple," sometimes called "plain," syrup. This was made by one of the following formulas.

SIMPLE SYRUP—COLD PROCESS

Granulated Sugar 12 pounds
Water 1 gallon

This will produce almost 2 gallons of simple syrup. Put sugar in a stone crock, pour on the water, and agitate with wooden paddle until sugar is entirely dissolved.

SIMPLE SYRUP BY PERCOLATION—COLD PROCESS

Sugar 12 pounds
Water 1 gallon

Put sugar into a large percolator, first putting some cotton in bottom of percolator so that sugar cannot run through (or barrel arranged as a percolator, by means of a perforated false bottom supported above the spigot). Add water, and let it drop slowly from bottom of percolator. When water has all run through it will have dissolved the sugar, and the syrup will be made. This process has the advantage of being automatic, the syrup making itself without work on the part of the dispenser.

SYRUP—HOT PROCESS

Sugar 12 pounds
Water 1 gallon
Put water and sugar in kettle, stir constantly as it warms, and take off the fire just as soon as it begins to boil, and strain while hot.

ROCK CANDY SYRUP

Sugar 16 pounds
Water 1 gallon
This will produce over 2 gallons of very heavy syrup. Put water and sugar in kettle, stir constantly until it boils up once; strain and set aside to cool. When it gets cold there will be a skin of crystallized sugar on the top; strain again to remove this skin, and you will have rock candy syrup, which is sugar syrup just below the crystallization point. You can reduce this with one-fifth of its bulk of water before using it to make fountain syrups.

Now, if our man chose *concentrated* syrups for flavoring his simple syrup, he used a one-to-three ratio, one quart concentrated lemon syrup, for example, to three quarts plain syrup. But if he had flavor *extracts* to choose from, the ratio was different, every time—as the extract table makes clear.

The extracts with asterisks are among those still available and suitable for making syrups. But certainly the least painful method is to begin with *fruit juices.* One-half pint of any fruit juice—cherry, cranberry, currant, grape, lemon, orange,

FLAVOR EXTRACTS

Add the number of ounces noted
to 1 gallon of simple syrup.
Extracts with asterisks are available today.

*Almond 1
Apple 1
*Apricot 1
*Banana 1
Birch Beer 2
*Blackberry 1
Blood Orange 2
Bouquet ½
Calisaya (Elixir) 4
*Celery 2
Champagne Cider 1½
*Cherry 1½
*Chocolate 3
*Cinnamon 1
*Coffee 4
*Cognac 1
Coriander 1
Cream Soda 2
Currant, Red 1
Claret 1
Don't Care 3
*Ginger, Jamaica 2
Ginger, Jamaica, Soluble 2
Ginger Ale, Dublin 1½
Ginger Ale, Hub ⅛
*Grape 1
Grape Cider 1
*Grenadine 2
Hoarhound 1½
Julep 1
Kola 2
*Lemon 1
Lemon, Sour 3
*Maple 2
Mead 2
*Nectar 1
Nerve Tonic 2½
*Nutmeg 1
Orange ½
*Orange, Florida 2
Orange Flower 2
Orange Cider 1
Orgeat 1
Orris Root 2
*Peach 1½
Peach Cider 1
Pear 1
Pear Cider 1½
Pepsin Soda 1½
*Peppermint 1
Phosphate Brain 1
*Pineapple 1
*Pistachio 1½
*Plum 2
*Raspberry 1
Raspberry Cider 1
Root Beer, Ottawa 4
Root Beer, Peruvian 4
*Root Beer 4
*Rose 2
Sarsaparilla 1
Sarsaparilla and Iron 2
Sherbet 1½
*Strawberry 1
Tonic Beer 1½
Wild Cherry 1
*Wintergreen ½
*Violet 1½

peach, pineapple—added to three pints of simple syrup, and the result: a manageable quantity of homemade syrup.

But if our man had high ideals—and plenty of time—he crowned that American merchandising marvel, "the weekly special," with his very own syrup. Not a syrup flavored with a concentrate, or an extract, or a fruit juice; but a syrup based on the flavor source itself: Take "green mint . . . three bunches. . . ." Below, culled from the pages of *How to Make a Soda Fountain Pay,* are recipes for some of those incredible syrups.

BANANA SYRUP

The syrup is prepared as follows: Cut the fruit in slices and place in a jar, cover with sugar and cover the jar, which is then enveloped in straw and placed in cold water, and the latter is heated to the boiling point. The jar is then removed, allowed to cool, and the juice is poured into small bottles. To 1 pint of this juice add 2 pints plain syrup.

CHOCOLATE SYRUP I

Powdered Cocoa 1 pound
Granulated Sugar 10 pounds
Water 1 gallon
Sweet Cream 6 ounces
Extract Vanilla ½ ounce
Put water on the fire with the cocoa and keep stirring to prevent lumping. When it comes to a boil, add the sugar and stir until it is all dissolved, then let it come to a boil and boil three minutes. Strain through cloth while hot. Allow to cool and then stir in the cream and vanilla extract.

CHOCOLATE SYRUP II

Powdered Chocolate 1 pound
Water 5 pints
Cornstarch 1½ ounces
Sugar 5 pounds
Dissolve the sugar in hot water, add the chocolate, using great

care not to scorch it. Mix the cornstarch with cold water, add to the chocolate, and boil for one minute over a slow fire, stirring constantly; strain through a sieve to remove coarser particles. A double boiler should be used, and burning thus avoided. Chocolates should always be boiled, for otherwise its full body and flavor cannot be obtained; therefore, be sure to bring your syrup to a boil. Cornstarch gives body, and, if properly cooked, is tasteless. If not thoroughly cooked, however, it gives a starchy taste, and, if burnt, its flavor is very noticeable.

Editors' Note: For a manageable quantity of homemade chocolate syrup, the modern soda jerk will find this adaptation of the cocoa recipe most suitable.

Sugar 2 cups
Cocoa, unsweetened ½ to ¾ cup
Salt ¼ teaspoon
Water 1 cup
Vanilla Extract ½ teaspoon
Heavy Cream 2 tablespoons

Combine sugar, cocoa, and salt; add water and stir to blend. Boil for five minutes, stirring constantly while the sugar dissolves. Cool, and then add vanilla extract and cream.

COFFEE SYRUP

Mocha Coffee ½ pound
Java Coffee ½ pound
Boiling Water 1 gallon

Boil together, or pass through a suitable filter coffee pot, until 1 gallon is obtained; settle, and add 10 pounds granulated sugar.

CREAM SYRUP

Sweet Cream 1 quart
Milk 1 quart
Sugar 4 pounds

Stir until sugar is dissolved.

HICKORY-NUT OR WALNUT-CREAM SYRUP

Take one pound of hickory-nut or walnut kernels and remove the skin by blanching—which, if left on, would give an unpleasant, bitter taste; then powder in a wedgwood or porcelain mortar, adding a few drops of lemon juice to prevent the separation of oil in kernels; water, gradually added, will make a thick emulsion. As fast as the kernels are reduced, put them in a linen cloth, which should be gathered around them so that they may be squeezed through the cloth. Whatever is left in the cloth should be returned to the mortar and pulverized further, the lemon juice and water being added as needed. All should eventually pass through the strainer. The result of this process (about two pints) is to be added to two quarts of cream syrup.

This formula may be varied, and perhaps improved, by a slight addition of extract of lemon or vanilla, or any other flavor to suit the taste; likewise a little coloring to suit the fancy. It will repay the labor of preparing it.

LEMON SYRUP I

Lemons, grated peel of 8
Lemons, expressed juice of 8
Granulated Sugar 6 pounds
Water 4 pints

Contuse the peel with 8 ounces of the sugar until the oil is thoroughly expressed and absorbed by the sugar, then mix with the remainder of the sugar, add the juice and water, dissolve by heat, strain, and cool.

LEMON SYRUP II

Lemons, juice and grated peel of 12
Sugar ½ pound
Syrup 1½ gallons

Grate the peel of the lemons and rub in a mortar with ½ pound sugar. Add the juice and let stand two hours. Strain through a cloth and wash with the syrup by straining through the same cloth. Agitate thoroughly.

MARMALADE SYRUP

Quince Syrup or Jelly 6 pints
Maple Syrup 2 pints
Mix and strain.

MARSHMALLOW SYRUP

One pound of white sugar, whites of 3 eggs, 6 ounces of gum arabic. Dissolve first the gum with 1 pint of water, strain through a fine sieve, cook the sugar with 1 pint of water to a thin syrup, let cool, beat the whites of eggs to a stiff froth; add the syrup by degrees, beating all the while, add the dissolved gum, beat until cold, put in a large glass jar, and cover air-tight. It is then ready for use. It is advisable to make only a small quantity in order to have it always fresh, because this syrup will sour in a few days.

MINT SYRUP

Green mint, 3 bunches; simple syrup, 6 pints; mix and bring to a boil over slow heat, strain through cloth and add simple syrup, 2 pints, and color light green.

ORGEAT SYRUP

Sweet Almonds (blanched) 8 ounces
Bitter Almonds (blanched) 3 ounces
Powdered Gum Arabic 3 ounces
Orange Flower Water 8 ounces
Water 3 pints
Rock Candy Syrup, enough to make 1 gallon
After the almonds have been blanched, beat them to a pulp in a mortar; add to this pulp the water, in small quantities at a time, stirring constantly and with pressure; strain through cloth. Dissolve in this the gum arabic, and add orange flower water. Lastly, add enough rock candy syrup to make one gallon.

VIOLET SYRUP

This preparation for soda fountains is made by adding a little strong tincture of orris-root to water, rendered clear, or nearly so, by filtration through magnesium carbonate, and dissolving in the flavored water enough sugar to make a syrup. Tincture of grass may be used as a coloring, if a green tint is desired.

REMEMBER ...

1 drop of extract to an ounce of syrup is equal to 2 drams to a gallon.
2 drops of extract to an ounce of syrup is equal to 4½ drams to a gallon.
3 drops of extract to an ounce of syrup is equal to 6½ drams to a gallon.
4 drops of extract to an ounce of syrup is equal to 1 ounce and 1 dram to a gallon.
5 drops of extract to an ounce of syrup is equal to 1 ounce and 3⅛ drams to a gallon.
6 drops of extract to an ounce of syrup is equal to 1 ounce and 5½ drams to a gallon.
7 drops of extract to an ounce of syrup is equal to 2 ounces to a gallon.
8 drops of extract to an ounce of syrup is equal to 2 ounces and 2½ drams to a gallon.
9 drops of extract to an ounce of syrup is equal to 2 ounces and 4½ drams to a gallon.
10 drops of extract to an ounce of syrup is equal to 2 ounces and 6¾ drams to a gallon.
12 drops of extract to an ounce of syrup is equal to 3 ounces and 3¼ drams to a gallon.
14 drops of extract to an ounce of syrup is equal to 4 ounces to a gallon.
16 drops of extract to an ounce of syrup is equal to 4 ounces and 4⅛ drams to a gallon.
18 drops of extract to an ounce of syrup is equal to 5 ounces and 1 dram to a gallon.
There are 450 drops to an ounce.

ICE CREAM,
ICES
AND SHERBET

Homemade ice cream! Yummy! The soda fountain hero of yesteryear turned his ice cream freezer by hand—a staggering amount of work. Toward the end, when the ice cream was almost frozen, it required sheer, brute force. But it was worth it: by the turn of the century, three out of every five concoctions served contained the ice cream that the modern cook—lucky soul!—can make in about half an hour in an electric ice cream freezer. For the recipes below, a 6 quart freezer is recommended; the yield, depending on the freezer, will be from 4 to 5 quarts.

VANILLA ICE CREAM I

A "no-cook" formula that goes from mixing bowl to ice cream freezer

Eggs, beaten 4
Sugar 2½ cups
Salt ½ teaspoon
Heavy Cream 4 cups
Milk 4 cups
Vanilla Extract 2 tablespoons

To the beaten eggs, slowly add the sugar and salt, stirring constantly until they are well mixed. Beat in the cream and the milk, stir in the vanilla extract, and turn into the ice cream freezer can. Lower the can into the freezer, position the dasher, and pack according to the freezer directions—usually 6 parts ice to 1 part rock salt in alternating layers. (Rock salt can be difficult to get in some areas, but Kosher cooking salt can be substituted.) Start the freezer going and in about half an hour, depending on your freezer, there it is—delicious homemade vanilla ice cream.

VANILLA ICE CREAM II

A formula heated over the stove for an ice cream with a delicious custard-like flavor

Milk 4 cups
Eggs, beaten 4
Sugar 2 cups
Salt ½ teaspoon
Heavy Cream 4 cups
Vanilla 2 tablespoons

Warm the milk and then add the eggs, sugar, and salt. Stir constantly until the mixture reaches a temperature of 155°—or until it coats the back of the spoon. Remove from heat, stir in the cream and vanilla extract, and continue stirring occasionally until the mixture has cooled to about 90°. (This will take about five minutes if the mix is transferred to a chilled bowl placed over ice.) Turn into the freezer can and freeze.

BANANA ICE CREAM

Follow either recipe above using just 1 tablespoon of vanilla extract. When the ice cream is frozen, remove the dasher and with a spoon move the ice cream to the sides of the can. Into the center dip 1 quart sliced and slightly mashed bananas that have been sweetened with ¼ cup sugar.

CHOCOLATE ICE CREAM

Follow either formula for ice cream, but use only ¼ teaspoon vanilla extract and 5 to 6 squares (5 to 6 ounces) unsweetened chocolate. The chocolate can be added to the "cook" formula along with eggs, sugar, and salt; it will melt as the warmed milk gets hotter. To add sugar to the "no-cook" formula, melt the squares of chocolate on the top of a double boiler, stirring constantly and taking care not to let it burn. When the chocolate is smooth, very slowly pour into its pan 2 cups of the ice cream mix, thoroughly stirring in each ¼ cup before adding more. Pour back into the ice cream mix, stir again, and turn into the freezer can.

And for a delightful chocolate chip "no-cook" ice cream, follow these directions: When the chocolate has melted, add 1 tablespoon to the cold ice cream mix and beat quickly; bits of chocolate will crystalize immediately and remain hard throughout the freezing process. Then dip out 2 cups of the ice cream mix and proceed as above, slowly stirring the mix into the chocolate.

CHRISTMAS EGG NOG ICE CREAM

To the second formula for ice cream, add 4 more eggs, stirring all the eggs, salt and sugar into the warmed milk. When the mixture has reached 155° (be *very* careful not to overheat), remove from heat and add the cream. While the mix cools, take a separate mixing bowl and in it combine thoroughly 10 tablespoons good brandy with 2 tablespoons vanilla extract, ¾ teaspoon nutmeg, and ¼ teaspoon cinnamon. Beat this into the mix and turn all into the freezer can. A delicious Christmas treat.

COFFEE ICE CREAM

Make ice cream with either formula but use no vanilla. Add instead 1½ cups of the very best, very strongest coffee you can make. For excellent results, use a filter coffee pot, pouring 2 cups of boiling water over 6 heaping tablespoons of the best espresso coffee.

PISTACHIO ICE CREAM

Make vanilla ice cream, but in place of 1 tablespoon vanilla extract use 1 tablespoon almond extract. Add 1½ to 2 cups of nearly pulverized pistachio nuts before freezing, and a drop or two of green food coloring.

STRAWBERRY ICE CREAM

Make ice cream but leave out the vanilla extract altogether. When the ice cream is frozen, remove the dasher and with a spoon move the ice cream to the sides of the can. Into the center dip 1 quart cleaned, partially mashed strawberries that have been sweetened with ¾ cup sugar.

The ice cream recipes above will give you indescribably luscious homemade ice cream. And now, look below for what to do next. These are the simple, old-fashioned toppings and combinations from *How to Make a Soda Fountain Pay.* They do not require soda, only ice cream in all its glory. Ice cream flavors are not specified, so use your imagination. With "Heavenly Twins," for example, try combining a scoop of banana ice cream with a scoop of "Pineapple Snow," made with the recipe in the sherbet and ice section that ends this chapter.

CANTALOUPE CREAM

Cut a small nutmeg cantaloupe in half. Remove the seeds and fill with ice cream. Serve with a spoon on thin china plate with maraschino cherries on each half. Of course cantaloupes should be kept ice cold.

CHERRY SUNDAE

Turn measure of ice cream in a champagne glass, pour over this several maraschino cherries and one ounce of cherry syrup. Serve with a spoon.

COLLEGE ICE
ALSO KNOWN AS "SUNDAE"

Into a flare champagne glass turn a cone-shaped measure of ice cream. Over this pour a half ounce of crushed fruit from bowl on counter, any flavor called for by the customer, who eats the ice cream, mixing it with the fruit as he proceeds, with delicious results. Chocolate Syrup over ice cream is also a favorite.

This method allows you to serve any flavor ice cream called for—but carry only vanilla in stock.

CREAM NECTAR FOR TWO

To 1 ounce sweet cream add 1 tablespoon powdered sugar and stir until dissolved. Add a teaspoon each of orange and pineapple syrups, a drop of vanilla extract, and a drop of red food coloring to color a pretty pink. Mix thoroughly and draw over ice cream. Serves 2.

HEAVENLY TWINS
VERY GOOD

Take 1 ladleful ice cream and 1 ladleful Pineapple Snow (see recipe below).
Top with whole strawberries or cherries. Serve in a sundae glass with a spoon.

HICKORY OR WALNUT SUNDAE
NEW AND VERY NICE

Into a champagne glass turn a cone of ice cream. Over top of this stick a few hickory or walnut meats broken into small pieces. Over all spoon an ounce of very strong apricot syrup. Serve with a spoon. Result is delicious beyond description.

ICE CREAM SANDWICHES

These have become very popular, and are made by spreading a thin portion of ice cream between two vanilla sugar wafers.

REMEMBER . . .

"Don't let the first impression you make be anything but a good one. Don't leave your white coat unbuttoned unless you have a white vest on, and don't lean on the counter."

MILK SHAKE

This beverage has attained great popularity. Wherever it has been properly introduced, it has immediately become extremely popular. It can be made of any flavor desired, but vanilla and chocolate are the most desirable.

This is the latest formula: Fill the tumbler half full of shaved ice, add 1½ ounces milk shake syrup (composed of equal parts of vanilla and pineapple syrups), fill tumbler almost full of milk, add 1 tablespoonful vanilla ice cream, and shake well.

A dash of port wine added to the milk shake syrup improves the drink.

MILK SHAKE: CHOCOLATE

Chocolate Syrup 1½ ounces
Milk 5 ounces
I tablespoonful of chocolate ice cream, and shaved ice to fill tumbler. Shake well, shave some ice on top, and serve.

MILK SHAKE: VANILLA

Fill the tumbler half full of milk; to this add 2 or 3 teaspoonfuls of sugar and a tablespoon of ice cream. Shake in a few drops of vanilla extract. Fill tumbler nearly full of finely broken ice. Shake vigorously until the ice is melted, place a long-handled spoon in the tumbler, and serve with two straws.

MINT SUNDAE

Fill saucer or glass nearly full ice cream, pour 1 to 2 ounces crême de menthe or crême de menthe syrup over the ice cream, and add 3 or 4 crême de menthe cherries. Serve in an ice cream saucer or a delicate champagne glass with small sherbet spoon and glass of water.

PINEAPPLE BONBON I
A SOUTHERN DAINTY

Fill a sherbet glass nearly full of finely shaved ice, add a ladleful of crushed pineapple from bowl on counter. Place a spoonful of ice cream on top, and serve with a sherbet spoon.

PINEAPPLE BONBON II

Take pineapple and a small quantity of each fruit in season. Cut very fine, place in a dish, add enough sugar to cover, and let stand five hours before using.

Ice Cream	1 large teaspoonful
Mixed Fruits	2 tablespoonfuls

Or, ladle into a glass, fill with soda water and serve with a spoon.

WALNUT SPONGE

Ice Cream, plain	1 large tablespoonful
Walnut Topping	½ ounce
Whipped Cream	2 large tablespoonfuls

Cover ice cream with walnut topping and add cream. Or, mix topping with half a glass of soda, add ice cream and cream, fill with soda and serve with a spoon.

ICES

The two recipes below are shortcuts to the ices popular at the turn of the century. They are the forerunners to the modern "snow cone"—and to learn how to make *real* ices, read on.

CLARET GLACE

Fill a champagne glass with finely shaved ice. Pour an ounce of claret syrup over the ice, and garnish with a thin slice of lemon.

CREAM CHERRIES

Fill sherbet glass almost full of finely chopped ice and add a ladleful of canned whole cherries. Top with spoonful of whipped cream, and serve with sherbet spoon.

The beautiful ices below are wonderfully easy to make. They are simple combinations of sugar, water, fruit and fruit juices, and for the best results should be turned into an ice cream freezer. The recipes will yield, depending on your freezer, from 2 to 3 quarts.

LEMON ICE

Water 4 cups
Sugar 2 cups
Lemon, juice of 10 (¾ cup)
Lemon rind, grated 1 tablespoon

Add the sugar to the water in a pan on the stove and stir until the mixture has reached a temperature of 180°—or until it is completely clear and close to boiling. Remove from the heat and add the lemon juice and rind, continuing to stir occasionally until the mix has cooled to about 90°. (This will take about 5 minutes if the mix is transferred to a chilled bowl placed over ice.) Turn into the freezer can, lower into the freezer, and pack according to directions. The result is a pale, pale lemon color, almost white. But it is refreshingly tart—and delicious.

ORANGE ICE

Defrost 1 can (6 ounces) frozen, concentrated orange juice. Follow the direction for lemon ice, but use the juice of only 2 lemons. Before turning the mix into the freezer can, stir in the orange juice and the grated rind of 1 large orange.

Orange ice was a great favorite at the turn of the century, and it was made in huge quantities. Sometimes it was served by itself, sometimes in sherbet glasses topped with sweetened strawberries, pineapple, or raspberries—and sometimes in a concoction called "Frozen Phosphate." *How to Make a Soda*

Fountain Pay carried this advice about the Frozen Phosphate: "As its name suggests, it is the coldest drink in the world, and is best served in the following flavors: orange, lemon, pineapple, raspberry, grape, cherry, claret, and chocolate. Draw into a glass 1 or 2 ounces of any of the above syrups, add a dash of phosphate (see Index), fill the glass almost full of soda, add a good-size scoop of orange ice, and serve with a straw. The orange ice, besides making the drink intensely cold, gives it a delicious flavor, as orange blends well with any of the flavors mentioned above."

PINEAPPLE ICE

Follow the directions for lemon ice, but use the juice of only two lemons. Before turning into the freezer, stir in 2 cups (1 can) of crushed pineapple in heavy syrup.

RASPBERRY ICE

Allow 2 cups sweetened frozen raspberries to defrost. Follow the directions for lemon ice, but use the juice of only 2 lemons, and before turning into the freezer, stir in the raspberries.

STRAWBERRY ICE

Follow the directions for lemon ice, but use the juice of 5 lemons. Before turning into the freezer, add 2 cups of fresh strawberries that have been sweetened with ¼ cup sugar. The strawberries can be mashed, or a few of them quartered and the balance blender pureed.

SHERBETS

Sherbets are as easy to make as ices, for they call for the addition of 1 cup milk to the basic ice formula. Again, these recipes will yield from 2 to 3 quarts.

FRAPPEED FRUIT

Orange Juice 1 can (6 ounces) frozen concentrated
Water 4 cups
Sugar 2¼ cups
Milk 1 cup
Lemons, juice of 3
Maraschino Cherries ¼ cup
Sliced Peaches ¼ cup
Whole Strawberries ¼ cup
Sliced Orange ¼ cup

Allow the frozen orange juice to defrost. Clean and slice the peaches and orange, add the cleaned, whole strawberries and cover all with ¼ cup sugar. Set aside. Add the balance of the sugar—2 cups—to the water in a pan on the stove. Stir until it has reached a temperature of 180°—or until the syrup is clear and close to boiling. Remove from heat and add the milk, orange juice, and lemon juice, stirring occasionally until the mixture has cooled to about 90°. (This will take about 5 minutes if it is transferred to a chilled bowl placed over ice.) Turn into the freezer can and when the sherbet is ready, remove the dasher and with a spoon move the sherbet to the sides of the can. Into the center dip the maraschino cherries and mixed fruit. To make orange ice, just leave out the cherries and mixed fruit and add the grated rind of 1 large orange.

GRAPE SHERBET

Grape Juice 1 can (6 ounces) frozen concentrated
Water 4 cups
Sugar 2 cups
Milk 1 cup
Lemon, juice of 3

Defrost the grape juice and combine the sugar and water according to the directions given in the above recipe; add the milk and lemon juice when the syrup is removed from the heat, stir in the grape juice, and cool as directed. Turn into the freezer can and freeze.

GRAPE AMBROSIA

Grape juice—and of course grape sherbet—was the great rage at the turn of the century. The juice was served in every conceivable way and grape sherbet appeared at soda fountain counters everywhere, always crowned with names as delicious as the sherbet itself. There was "Grape Juice Frappe" and "Grape Chill" and "Grapes of Ice." One fountain owner beat the whites of two eggs with a tablespoon of sugar and added the froth to the grape sherbet just before turning the mix into the freezer. That was dubbed "Iced Grapes." Another added grated orange peel, and still a third added pineapple and called it "Grape Ambrosia." Make grape sherbet and when it is frozen, remove the dasher and add 1 cup diced fresh pineapple sweetened to taste with about ¼ cup sugar; or, stir in crushed canned pineapple, the juice drained off.

PINEAPPLE SNOW

Follow the recipe for basic sherbet, using the juice of 2 lemons. Before turning into the freezer, add 2 cups of canned, crushed pineapple in heavy syrup and 1 tablespoon of finely grated lemon peel.

STRAWBERRY SHERBET

Follow the recipe for basic sherbet, using a total of 5 lemons. Before turning into the freezer can, stir in 2 cups of fresh strawberries that have been sweetened with ¼ cup sugar. The strawberries can be mashed, or a few quartered and the balance blender pureed.

REMEMBER . . .

"Do you care to drink water that comes from a common sidewalk fountain? And yet it's pure water! What wonder, then, that people patronize a polished and shining soda fountain rather than an untidy and tarnished one?"

"About ice cream sodas: The best ice cream
will bring the best trade both in the class of
customer attracted and in the volume of business done.
No artificial thickener can give
the smooth, rich flavor of pure frozen cream. Also use nothing
but the very best vanilla extract
in flavoring ice cream. If artificial or cheap
vanilla extract is used, it is bound to show.
In the matter of ice cream for sodas,
the best *is* the cheapest, also very much the best."

The ice cream freezer

FIZZES, SODAS AND FLIPS

The nearly fifty recipes that follow are for drinks that enjoyed moments of stardom as the twentieth century rolled on its way. The few exotic ingredients are explained in notes, but the great majority are available in today's supermarkets. The drinks are thus easy to make, easy to double (or to multiply indefinitely), and they are, obviously, delicious. Note that nearly all of them call for 1 to 3 ounces of syrup—proportions that the soda jerk varied according to taste. And remember that while ice cream is rarely called for, it was almost always added. So, mix your syrups, fill your glass half full of club soda, and stir; add your favorite ice cream, fill to the top with club soda, and serve.

AMYCOSE

Shaved Ice	½ tumblerful
Raspberry Juice	1 ounce
Orange Syrup	2 ounces
Orange, juice of	½

Shake well, add soda water, and before serving add a small, thin slice of orange or pineapple. Serve with two straws.

ANGEL WING
SOMETHING NEW

Draw 1½ ounces of crême de menthe or crême de menthe syrup in a glass, fill within an inch of the top with soda, stir with a spoon, then float on top of the drink an ounce of cream. Serve without a straw.

BANANA FLIP

Shaved Ice	½ tumblerful
Banana Syrup	2 ounces
Cream or Milk	8 ounces

Shake well, add a few small pieces of banana and fill with soda. Serve with spoon and straw. This is a seller.

Editors' Note: A recipe for homemade banana syrup appears in the chapter on syrups, or vanilla syrup with crushed banana may be substituted.

BROWNIES

One ounce kola syrup in a mineral glass, add dash of ginger extract, soda, stir, and serve.

BUFFALO

Made with 11 ingredients and sells for 11 cents. Ice, vanilla, nectar syrup, lemon juice, sherbet, grape juice, sprig of mint, slice of pineapple, slice of orange, strawberries and a maraschino cherry!

Mix 1 ounce nectar syrup with a dash each of vanilla extract, lemon juice, and grape juice. Fill glass half full with chipped ice and soda, add a scoop of lemon sherbet, and fill with soda. Top with fresh fruit and mint.

REMEMBER . . .

"Women like a little more syrup and three times as much foam and froth as the men like. But never overflow the glass."

CATAWBA COBBLER

Fill a glass ⅓ full of fine ice. Add 1 ounce lemon syrup, 1 ounce grape juice, and then fill with soda. Serve with straws; top with slice of orange.

CHARLOTTE RUSSE
A LADIES' DRINK

Put together 1½ to 2 ounces of vanilla syrup and a little ice cream. Shake well in shaker, pour into glass and fill nearly full with soda water, then fill with whipped cream, flavored with vanilla, and serve with spoon.

CHOCOLATE BOUQUET

Mix 1 ounce chocolate syrup and 1 ounce orange syrup in a large soda glass. Add scoop of ice cream, and draw soda to fill glass.

CLARET LIMEADE

Squeeze 2 limes into a glass and stir in 3 teaspoonfuls of powdered sugar until dissolved. Add 1 spoonful of shaved ice, and fill with soda. Pour an ounce or two of claret wine over the top and serve with a straw.

CLARIQUE

Shaved Ice ½ tumblerful
Claret 2 ounces
Lemon Syrup 1 ounce
Powdered Sugar 3 teaspoonfuls
Dissolve sugar in syrup and claret, add ice, draw soda, and place a few strawberries on top.

CLEVELAND GEM

Serve in a glass half full of cracked ice. Add ½ ounce each of raspberry, currant, lemon, and claret syrups, and 1 ounce plain cream. Fill glass with soda. Top with nutmeg. Serve with straws.

COKA KOLA

Glasses should be thin, and hold not over 8 ounces. Draw only 1 ounce of kola syrup to the glass (one part of kola to seven parts of soda water, and no stronger). Then fill the glass with soda, and stir thoroughly with a spoon. A dash of celery extract greatly improves this drink.

CREAM CITY

Fill shaker half full of cracked ice; add 1 ounce chocolate syrup, 1 ounce strawberry syrup, 2 ounces plain cream, 1 tablespoonful ice cream; shake, and transfer to glass without straining. Fill with soda and serve with straws.

CRUSHED VIOLETS

Draw 2 ounces violet syrup in a soda glass, add a scoop of ice cream and fill with soda. Sprinkle on top of foam a few candied violet leaves.

Editors' Note: Violet syrup can be made with violet flavor (Flavor Mill), and, if a gallon sounds like a handy amount, follow the instructions given in the extract table in the first chapter. Or, dissolve 2 tablespoons of confectioners sugar in a little milk and flavor with the violet flavoring. Add ice cream and soda and shake well.

CUBANADE

Orange Syrup 1 ounce
Grape Juice 1 ounce
Lemon, juice of ½
Jamaica Ginger from shaker A pinch
Shaved Ice ¼ glass
Mix thoroughly, fill with soda and garnish with a sprig of mint and a maraschino cherry.

FROSTED COFFEE

Fill shaker half full of cracked ice. Add 2 ounces coffee syrup, 2 ounces plain cream, 1 tablespoonful ice cream, shake, fill shaker with soda and strain; serve with straws.

FRUITINE

Dissolve two heaping teaspoonfuls powdered sugar in juice of ½ lemon and ½ orange. Fill glass ⅓ full of cracked ice, add 4 strawberries, a slice of pineapple and 2 maraschino cherries. Add soda to fill glass and garnish with a sprig of mint.

GINGER ALE COBBLER

Sugar one lump, one slice pineapple, dash of angostura bitters, ice; fill glass with ginger ale, and stir.

GINGER COCKTAIL

Mix ½ ounce of grape syrup and 3 ounces orange juice and fill glass with ginger ale. Add shaved ice and serve.

REMEMBER . . .

"If you want the best of everything, stand your tumblers with the rims in crushed ice. They will retain their coldness a long time. It pays to study these minor details."

GOLF GOBLET

Fill glass ⅔ full of cracked ice, add ½ ounce pineapple juice, 1 teaspoonful lemon juice, 1 teaspoonful Raspberry Vinegar. Put spoon in glass, and fill to within ½ inch of top with soda; add shaved ice heaping full. Put strawberry or cherry on top and stick slice of orange down side of glass. Serve with spoon and straws.

Editors' Note: "Raspberry Vinegar" is no longer available, but a dash of 4 Fruit Vinegar or Pear Vinegar (Harry and David) may be added to a teaspoon of raspberry syrup. Or, simply flavor the raspberry syrup with a drop of cider vinegar.

GRAPE CHILL

Partly fill a thin glass with cracked ice, and pour in as much grape juice as the glass will accommodate. Serve with or without a straw. This is a most satisfactory drink.

HONOLULU PUNCH

Into soda glass draw ½ ounce each of pineapple and strawberry syrup, 2 ounces claret wine and a spoonful of shaved ice. Fill glass with soda; stir. Top with maraschino cherry and sprig of fresh mint.

HORSE'S NECK
LONG AND COOLING

Preferably served in a stem glass. This is a favorite in New York City during the Horse Show.

Fill glass ⅓ full cracked ice; add 1 dash orange bitters. Fill glass with ginger ale. Stir; lay a slice of lemon on top of glass, and serve with a straw through the lemon slice.

IMPERIAL CREAM

Extract Vanilla 1 dash
Pure Cream 1 ounce
Shaved Ice ½ tumblerful
Egg, white of 1
Shake well, strain into a glass and fill with cream soda.
Sprinkle nutmeg on top and serve.

JACQUEMINOT

Shaved Ice ½ tumblerful
Rose Syrup 2 ounces
Milk 8 ounces
Egg, white of 1
Shake well, add soda water, and serve with straws.

Editors' Note: Substitute a dash to taste of rose flavor (Flavor Mill) for the syrup. This can be added to 2 tablespoons of confectioners sugar dissolved in a tablespoon of the milk.

LADIES' COCKTAIL

In a mixing glass half full of cracked ice, add 2 ounces cherry syrup, one dash lemon juice, 1 teaspoonful brandy, and fill with plain water. Strain into cocktail glass and add maraschino cherry.

LEMONADE PLAIN

Lemon, juice of 1
Cracked Ice ½ tumblerful
Powdered Sugar 1 tablespoonful
Ice Water 8 ounces
Dissolve sugar in lemon juice, add ice and water, shake thoroughly, then add slices of lemon and orange.
Serve with 2 straws.

LEMONADE SODA

Lemon, juice of 1
Cracked Ice ½ tumblerful
Powdered Sugar 1 tablespoonful
Dissolve sugar in lemon juice, add ice, shake thoroughly, then add soda to fill glass. Top with slice of lemon and a maraschino cherry, and serve with a straw.

LIMEADE

Dissolve 3 teaspoonfuls of powdered sugar in the juice of 3 limes—not lemons; add cracked ice and fill with soda. The limes may be bought by the barrel at a low price; and, as the fruit is small (not much larger than a walnut), this item is of trifling expense.

MAGDA

An uptown druggist, says the New York *Sun,* has added another laurel to his fame. A blonde-haired young woman came into his store the other day, and stepped up to the soda water counter. "I'll take a Magda, with ice cream, please," she said; and the man in charge mixed a glass of soda which looked, when it was ready, like ordinary chocolate soda.

An undecided man standing near the young woman waited until the drink had been consumed, and then asked the druggist what it was.

"Oh, that's chocolate and coffee syrup mixed," he said, "and I call it 'Magda.' That's one of the things Magda did in the play that shocked her family when she came back as a famous singer—drink chocolate and coffee mixed. I saw Duse in the play, and read that in the libretto; and I made up my mind then that this summer I should have a new kind of soda, and call it 'Magda.' I put a sign in the window for a few days, and already it is one of the drinks that people ask for most. I don't know whether or not they all know where the idea came from, but they seem to like the result just as well as though they did." This illustrates the value of keeping abreast of the times. A dispenser by occasionally putting on some new combination and calling it after an event or book or play that is at the time creating a furore, can advertise and make popular his soda fountain and increase his business.

A MODERN MAGDA

Chocolate Syrup 1 ounce
Coffee Syrup 1 ounce
Ice Cream 1 scoop
Combine the syrups in a tumbler, fill half full with club soda and stir. Add ice cream and fill to brim wtih soda. Sprinkle top with shaved bitter chocolate, and serve with straw and spoon.

MINT COCOA

To a shaker ⅓ full cracked ice, add 1 ounce crême de menthe or crême de menthe syrup, 1 ounce chocolate syrup, 2 ounces plain cream, and white of one egg. Shake. Fill shaker with soda; strain into glass. Serve with straws.

MINT FREEZE OR IRISH ICEBERG

Fill a champagne glass with finely shaved ice. Over this pour 1 ounce of crême de menthe or crême de menthe syrup, and finish with a thin slice of lemon and a crême de menthe cherry.

MINT LEMONADE

Add to your regular formula for lemonade 1 to 2 ounces crême de menthe or crême de menthe syrup. Strain into glass. Trim with slice of fruit and fresh mint and a crême de menthe cherry. Serve with straws.

REMEMBER . . .

"Chairs are a good thing: when resting, customers drink slowly. This means looking about and finding other things to buy."

MOORISH SHERBET

Shaved Ice ½ tumblerful
Strawberry Syrup ½ ounce
Pineapple Syrup ½ ounce
Vanilla Syrup ½ ounce
Milk sufficient to fill your 14-ounce tumbler.
Shake well, add soda, and pour from tumbler to shaker several times. Serve with straws.

OLIVE COCKTAIL

Soda water drinkers are always looking for something new. Olives have become very popular now in cocktails, and they can be used as a novelty with the following formula. Draw into a soda tumbler:
Champagne Syrup 1 ounce
Raspberry Syrup ½ ounce
Add a little shaved ice and a dash of angostura bitters. Mix well with a spoon, strain into a champagne glass, and fill with soda. Stick stuffed olive on a wooden toothpick and float in the glass. Twist a piece of lemon peel over top of the drink and serve quickly.

Editors' Note: Champagne syrup has disappeared forever, but grownups are advised that ½ ounce raspberry syrup has often graced a glass of champagne. History does not record the presence of the olive.

PEACH COBBLER

Peach Syrup 2 ounces
Whipped Cream ½ ounce
Plain Cream 1 ounce
Shaved Ice ¼ tumblerful
Draw peach syrup and plain cream in shaker with ice, shake well, fill with fine stream soda, add peach slices and whipped cream on top.

PEACH CREAM

Peach Syrup 1 ounce
Plain or Whipped Cream 1 ounce
Serve in a glass filled with soda and topped with grated nutmeg.

PERSIAN SHERBET

Shave ice into a shaker and add 2 ounces strawberry syrup. Fill nearly full with soda and add a few drops of lemon extract and orange extract (colored bright red). Shake very hard and serve in a glass.

PINEAPPLE SMASH

Pineapple Syrup 1 ounce
Shaved Ice ½ tumblerful
Powdered Sugar 1 teaspoonful
Dissolve sugar in syrup and add soda water. Strain in small tumbler, and add a piece of pineapple.

QUEEN'S FAVORITE

Into a glass filled ⅔ full of shaved ice, pour orange syrup 1 ounce, grape juice 1 ounce, dash of lemon juice and Jamaica ginger; fill with soda and serve.

RASPBERRY ADE

Raspberry Syrup 1 ounce
Lemon, juice of ¼
Add shaved ice and fill glass with soda. Serve with a straw.

REMEMBER . . .

"Don't argue with a cranky customer—satisfy him. Don't hesitate to change a drink which is not appreciated. And don't keep a customer waiting for a check."

RASPBERRY AND HONEY

Maple Syrup
Honey
Raspberry Juice
Draw ½ ounce of each of the above and fill tumbler with soda water, drawing quietly. Stir gently with a spoon.

ROOT BEER CREAM

Fill a glass within 1 inch of the top with root beer. Then float on top about 2 ounces of sweet cream. Do not allow your customer to use straw, and you have won his patronage on root beer forever.

ROSE MINT

Rose Syrup ½ ounce
Mint Syrup ½ ounce
Cream 3 ounces
Mix, fill with soda, and serve with shaved ice and straws.

Editors' Note: As a substitute for rose syrup, add a dash of rose flavor (Flavor Mill) to 1 ounce of the mint syrup.

SPA FIZZ

Shaved Ice ½ tumblerful
Lemon, juice of ½
Strawberry Syrup 2 ounces
Orange Syrup 1 ounce
Shake well. Draw soda water and serve.

WHITE MOUNTAIN FOAM

Shaved Ice ½ tumblerful
Cream ¼ tumblerful
Egg, white of 1
Orange Syrup 1 ounce
Draw soda water, stir, and serve.

NOSTRUMS, EGG DRINKS AND PHOSPHATES

The potions with which our forebears aided their digestions, recovered from hangovers, and generally rid themselves of the blahs—these were nostrums. "Ammonia Cocktail," for example; recommended as a "great bracer" and containing two dashes of aromatic spirits of ammonia. Not quite enough to do you in, but close.

Equal amounts of malt nutrine and simple syrup in a glass of soda was the specialty for bicyclists: "At once refreshing and nourishing." There was nerve tonic extract in soothing combination with simple syrup. And there was grape juice, one of the great cure-alls of the age. "A most healthful, delicious, and nourishing beverage," the dispenser book announced; "a delicate liquid food, keenly relished, and readily assimilated without taxing the digestive organs. Recommended by physicians for the sick and convalescent."

Headache? Bromoseltzer was the answer, fizzing up through a mint-topped lemon soda. "Prairie Oyster" started the day off with a bang: "Draw dash of soda water in a glass, add one egg, a splash of vinegar, season with pepper, salt, lemon juice, and serve." For the faint of heart, a glass of good old "2¢ plain" was held at the ready.

Oysters themselves were—and still are—considered a stoutly healthy food. They were served as nostrums with this recipe: "Mix and beat well together a tablespoonful each of good tomato ketchup and of strained lemon juice, 7 drops of Tabasco sauce, ½ teaspoonful of Worcestershire sauce, a pinch of paprika and the same of salt, and lastly a teaspoonful of horse-radish. Have ready in small glasses 30 small oysters, which have been on ice for one hour. Divide them among the sauce, which has also been set on the ice until very cold. Put five oysters in each glass. Serves 6."

Nostrums are gone from the soda fountain counter, and so are real egg creams. For, once upon a time, an egg cream had somewhere in its luscious depths an egg . . . *and* a dollop of cream. No more. Search in vain for an egg cream worthy of its name. Search in vain, for that matter, for a phosphate.

EGG DRINKS

When you make the following drinks, remember that if you do not serve the drink immediately, the shaker or glass must be covered tightly with plastic wrap and refrigerated. Foods containing raw eggs must *never* be allowed to sit in the open air.

And here, from *How to Make a Soda Fountain Pay,* further advice about egg drinks. "To make them properly, more care and attention should be given to the mixing of egg drinks than any other kind. And it pays to give them a good deal of attention, for five cents extra can be asked for the addition of an egg in the drink. But customers are more critical about egg drinks than any other kind. To be good, they must be properly made. In making an egg drink, always break the egg into the glass first, so that the customer can see it. An egg may be cracked neatly by holding it in the right hand between the thumb and forefinger, and striking it once on the edge of the glass which is held in the left hand. Immediately place the egg over the glass and gently press the thumb and finger together, which will divide the shell, and the egg will drop out. The yolk may be kept from breaking by letting the egg drop on the side of the glass. Do not break the egg into the shaker. Let the customer see the fresh egg to give him confidence and cause expectation of a good drink. Add the syrup and other ingredients, and two or three small pieces of ice to break up the egg thoroughly while shaking. The use of separate shaker and strainer is recommended for convenience. Set the glass on the counter, and, holding the shaker firmly in the right hand, place it over the glass, picking up the latter with the left hand. Shake from side to side quickly and vigorously. Leave the mixture in the shaker and draw the soda. Then pour into the glass, holding the shaker close to it, which leaves a fine foam on top. A slight sprinkling of finely ground nutmeg or cinnamon may be added as a top dressing, if desired. When lemon flavor is used, the rind of a piece of lemon should be twisted over the top, to add a nice fresh lemon flavor. The drinks will pour and have a finer foam at last, if the strainer is used. This will also take out any small pieces of ice which may not have melted. When it is desired that ice be left in the drink, serve with straws."

CARNATION FLIP

To a fresh egg, add ½ ounce each of pineapple, strawberry, and raspberry syrups, a dollop of cream and a teaspoonful of ice cream. Stir, add cracked ice, shake, fill with soda and serve. Sprinkle a little nutmeg on top. Serve immediately.

CATAWBA FLIP

Ice Cream 1 ounce
Egg 1
Grape Syrup 1 ounce
Chipped Ice A sufficient quantity

Put all into a shaker, and mix thoroughly. Strain into a glass, fill with soda water, dust some nutmeg over the top, and serve immediately. Sells like hot cakes.

CHARLOTTE RUSSE

Shaved Ice ½ tumblerful
Vanilla Syrup 1 ounce
Cream 6 ounces
Egg 1
Shake well and fill with soda water. Serve immediately in a tumbler with a spoon.

CHERRY BOUNCE

To a shaker with small amount of cracked ice, add 1½ ounces cherry syrup, 2 ounces plain cream, ¼ ounce brandy, 1 dash angostura bitters, 1 egg. Shake, strain, throw ice away. Fill glass nearly full of soda, pour quickly to shaker and back. Top with nutmeg, and serve immediately with a straw.

CIDER EGG NOG

Shaved Ice ¼ tumblerful
Cider 6 ounces
Vanilla Syrup 2 ounces
Egg 1
Shake well, strain through a julep strainer into a tumbler, and serve immediately with straws.

COFFEE NOG

Egg 1
Shaved Ice ½ tumblerful
Coffee Syrup 1½ ounces
Port Wine 1 ounce
Milk 6 ounces
Shake well, and fill tumbler with soda water. Serve immediately with a spoon.

CREAM FLIP

Fresh Egg 1
Fresh Cream 1 ounce
Vanilla Syrup 2 ounces
Crême de Menthe or Crême de Menthe Syrup ½ ounce
Mix thoroughly, add ice cream, and fill with soda.
Garnish top with nutmeg and serve immediately.

CREAM PUFF

This is one of the finest egg drinks made. Break a fresh egg into a shaker, draw an ounce of orange syrup, add a good-sized spoonful of ice cream, and shake very thoroughly. Then, without straining, fill the shaker with soda. Pour from shaker to glass, top with nutmeg, and serve immediately with a straw.

Chocolate Cream Puff and Coffee Cream Puff are also very fine, and can be made by using syrups named.

EGG ADE

Muddle ½ ounce brandy together with a teaspoonful of powdered sugar. Fill glass half full of cracked ice, add 1 egg and a dash of angostura bitters. Shake well, then fill glass with ginger ale, and serve immediately with a straw.

EGG ARMOUR

Chocolate Syrup 2 ounces
Egg 1
Sweet Cream 2 ounces
Cracked Ice A little
Malted Milk 1 teaspoonful
Shake well, add soda to fill glass, sprinkle with small quantity of salt, and serve immediately with a straw.

EGG COCOA

Take 1 teaspoonful of powdered cocoa, dissolve it in ¼ of a glass of hot water, add 2 or 3 spoonfuls of sugar, according to taste, put in 1 egg, and shake it together with cracked ice. Serve immediately. Costs about 2½ cents, sells for 10 cents.

EGG COFFEE

Shaved Ice ½ tumblerful
Fresh Egg 1
Coffee Syrup 1½ ounces
Cream (genuine) 2¼ ounces
Spring Water 5 ounces
Shake thoroughly. Draw small quantity soda water and serve immediately. The substitution of chocolate syrup for the coffee gives your customer an excellent Egg Chocolate.

EGG CREAM

Fresh Egg 1
Vanilla Syrup 2 ounces
Whipped Cream 1 ounce
Sherry Wine A dash
Fill with soda, dust with nutmeg, and serve immediately. The use of whipped cream makes this a specially smooth and delicious drink.

EGG LEMONADE

Shaved Ice ½ tumblerful
Fresh Egg 1
Juice of 1 large lemon
Powdered Sugar 3 heaping teaspoonfuls
Spring Water 6 ounces
Shake thoroughly. Draw small quantity soda water, and dust a little nutmeg on top. Serve immediately.

EGG MILK

Shaved Ice ½ tumblerful
Fresh Egg 1
Milk 8 ounces
Sugar (powdered) 3 teaspoonfuls
Shake thoroughly. Draw small quantity of soda water, and dust a little nutmeg on top. Serve immediately.

EGG TONIC

Fresh Egg 1
Orange Syrup 1 ounce
Angostura Bitters A dash
Shake well and fill with ginger ale. Serve immediately.

FLOWING STREAM

Shaved Ice ½ tumblerful
Egg 1
Vanilla and Orange Syrup 1 ounce each
Ice Cream 1 tablespoonful
Flavor the combined syrups with a dash each of angostura bitters, rum, and orange extract. Add ice cream, shake well, and fill with a small quantity of soda water. Serve immediately with spoon and straws.

GOLDENADE

Shaved Ice ½ tumblerful
Powdered Sugar 4 teaspoonfuls
Lemon, juice of 1
Egg, yolk of 1
Shake well, add soda water, turn from tumbler to shaker and *vice-versa* several times, and strain through julep strainer into a glass. Substitute the white of an egg for the yolk and you have the delicious Silverade. Serve immediately.

LIME JUICE SHAKE

Egg 1
Lime Juice ½ ounce
Lemon Syrup 1 ounce
Shaved Ice ¼ tumblerful
Shake well, fill glass with plain soda, pour from shaker to glass and serve immediately with 2 straws.

ORGEAT AND EGG

Shaved Ice ½ tumblerful
Orgeat Syrup 2 ounces
Egg 1
Shake well; strain through a julep strainer into a tumbler; fill tumbler with soda water, and serve immediately with straws.

PISTACHIO PUNCH

Draw 1 ounce lemon syrup into a tumbler; add ¼ ounce pistachio syrup, ½ ounce milk, ½ ounce claret wine, 1 egg, and shaved ice; shake thoroughly, strain, fill with soda, and top off with nutmeg. Serve immediately.

QUEEN'S FAVORITE

Shaved Ice ½ tumblerful
Pineapple, Raspberry, and Vanilla Syrups ½ ounce each
Egg 1
Add enough milk to nearly fill a glass. Shake well, fill glass with soda water, dust nutmeg on top, and serve immediately.

RASPBERRY PUNCH

Shaved Ice ½ tumblerful
Egg 1
Raspberry Syrup 2 ounces
A few squirts of rum and a drop of orange extract, colored red

with ruby, milk enough to fill glass nearly full. Shake well, add small amount of soda water, and dust a little nutmeg on top. Serve immediately.

SQUARE MEAL
A VERY HEARTY DRINK

Draw 2 ounces of chocolate syrup into a shaker, break in a fresh egg, add a spoonful of ice cream, and fill shaker with milk. Shake thoroughly, pour into a glass and top with nutmeg. Serve immediately.

PHOSPHATES

"Phosphates can be made of all flavors; but fruit phosphates are the most delicious and palatable, possessing decided and exquisite flavors. They satisfy the thirst, and are delightful and pleasing drinks, producing the most refreshing and invigorating effect upon the system.

"The best way to draw a phosphate is to draw soda in a glass within an inch of the top, then draw 1 ounce of the syrup desired, add a dash of phosphate, pour into another glass and serve. There are, however, many other ways all good, such as drawing desired syrup in a glass, adding dash of phosphate, filling with soda and stirring with spoon. Some people serve them with shaved ice. Phosphates are cooling, refreshing, palatable, and very profitable, and when good are one of the mainstays of a soda business."

For the modern cook, the following formula for phosphate solution is recommended. Crush or grind sour salt crystals (Wagner's or McCormick's) to a powder and dissolve in cold water to taste—at least ⅛ teaspoon to 2 cups water. The phosphate can be stored in a covered jar in the refrigerator.

REMEMBER . . .

"The sound of a fiddle makes a man want to dance, the odor of beefsteak makes him hungry; the sight of a tempting soda awakens his thirst."

BLACKBERRY BLOOM
A SPLENDID AUGUST DRINK

Draw 1½ ounces blackberry syrup into a mineral glass, add a dash of phosphate, half fill the glass with shaved ice and then fill with soda, top with a thin slice of lemon, and serve with a straw.

CHERADE

One ounce cherry syrup, ½ ounce orange syrup, a dash of phosphate, a dash of angostura bitters in a glass; fill with soda. Add a little shaved ice, 1 whole cherry, and serve with a straw.

CHERRY CELERY

Draw 1 ounce cherry syrup into a phosphate glass. Add a dash of celery extract, a dash of phosphate, fill with soda, stir and serve.

CHERRY COCKTAIL

Into a phosphate or champagne glass draw ½ ounce each of claret syrup and cherry syrup. Add a little shaved ice, a dash of phosphate, a dash of angostura bitters, and fill with soda.

Stick a maraschino cherry on the end of a wooden toothpick and float in the glass. Customer drinks, then eats cherry by the aid of the toothpick.

CHERRY GLACE

Take a small sherbet glass, fill with shaved ice, add 1 ounce cherry syrup and a dash of phosphate, fill with soda, and top off with 3 or 4 maraschino cherries. This makes one of the nicest drinks for a hot day that can be served at the fountain.

CHERRY PHOSPHATE

Dispense 1½ ounces cherry syrup, a dash of phosphate, and a liberal amount of soda water. The result will be delicious.

CLARET FRANCAISE

Draw 1 ounce each claret wine and strawberry syrup in a mineral glass. Add dash of phosphate, fill glass with soda, stir and serve.

CLARET MINT
TRY THIS ONE

Draw ½ ounce claret wine and 2 ounces crême de menthe or crême de menthe syrup in a mineral glass, add dash of phosphate, fill glass with solid soda, add a spoonful of cracked ice, garnish with a maraschino and a crême de menthe cherry.

CLARET PHOSPHATE

Draw 1 ounce claret syrup in a mineral glass, add dash of phosphate, fill with soda, stir and serve.

CRANBERRY PHOSPHATE

Add a dash of phosphate to a glass of cranberry juice and fill with ice.

EGG PHOSPHATE

Pour some cracked ice into a shaker, break in a fresh egg, add an ounce of orange syrup and a dash of phosphate. Shake well, then strain into soda glass. Draw fine stream to make the drink creamy, then pour back and forth from shaker to glass. Sprinkle top with nutmeg, and serve with a straw. Claret, lemon, and chocolate syrups may be substituted, or combine lemon with coffee, chocolate with orange.

GRAPE CHAMPAGNE

Grape Syrup 1 ounce
A dash of Brandy
Add a dash of phosphate, fill with soda and serve.

GRAPE PHOSPHATE

Grape Syrup 1½ ounces
Phosphate A sufficient quantity
Soda Water To fill glass
Tastes vary greatly as to the quantity of phosphate to be added; some like only a little, while others like it quite acid. The taste of the customer should be consulted. This drink should always be served solid in a small to medium glass. It is the greatest "thirst quencher" known, and is very popular.

LEMON AND LIME

Lime, juice of 1
Lemon Syrup ½ ounce
Few dashes phosphate, ¼ glass shaved ice. Draw soda water in mineral tumbler, and serve. This makes a fine morning drink.

LEMON GRAPE PHOS

Grape Syrup ½ ounce
Lemon Syrup ½ ounce
Add a dash of phosphate, fill with soda, and stir gently with spoon. Orange syrup may be substituted for the lemon with equally delicious results.

REMEMBER . . .

"Soda water is a matter of taste—in a double sense. Scores of customers lose their thirst when they see a slovenly fountain. Keep your fountain clean."

MINT EGG PHOS

Crême de Menthe or Crême de Menthe Syrup 2 ounces
Egg 1
Cracked Ice ¼ tumblerful
Whipped Cream 1 ounce
Shake well, fill shaker with soda, add a dash of phosphate, pour in a glass and sprinkle with nutmeg.

MINT JULEP

Crême de Menthe or Crême de Menthe Syrup 1 ounce
Phosphate 1 dash
Shaved Ice ½ tumblerful
Whiskey 1 tablespoonful
Slice of orange; soda to fill glass; top with sprig of mint and a crême de menthe cherry.

MINT PHOS

Crême de Menthe or Crême de Menthe Syrup 1 ounce
Phosphate 1 dash
Soda to fill glass. Stir and serve.

NECTAR
A LADIES' PUNCH

Lemon Syrup ½ ounce
Strawberry Syrup ½ ounce
Orange Syrup ½ ounce
Phosphate A dash
Orange, slice of 1
Add shaved ice, fill glass with soda, top with maraschino cherries, and serve with a straw.

ORANGE PHOSPHATE

Orange or Orangeade Syrup   1 ounce
Phosphate   1 dash
Serve with soda to fill glass.

SHERRY COBBLER

Muddle equal quantities of good sherry and powdered sugar with juice of fresh lime or lemon. Serve 2 ounces to the glass with a dash of prosphate. Fill with soda and shaved ice, stir, and top with slice of lemon.

SIBERIAN FLIP

Orange Syrup   1 ounce
Pineapple Syrup   1 ounce
Angostura Bitters   2 drops
Phosphate   2 dashes
Cream   1 ounce
Soda to fill glass. Top with very thin slices of orange and pineapple and serve with a straw.

WILD STRAWBERRY PHOSPHATE
A PRIZE WINNER

Strawberry Syrup   1 ounce
Lemon Syrup   1 ounce
Cherry Extract   A dash
Add a dash of phosphate, fill with soda and serve.

REMEMBER . . .

"You serve that drink which has made the people of America nearly forego the ale of England, the beer of Germany, the wine of France, and of which Samuel Johnson would have surely said, had he lived in our time, 'The cup that cheers, but not inebriates.'"

PUNCHES, SUPER WHAMMIES AND MODERN PARTY DESSERTS

The following recipes were among the prized possessions of the soda fountain hero: recipes for glorious super whammies, splendid punches, and simple, reliable formulas for large quantities of the great favorites—lemonade, iced tea, orangeade. And, if you have planned a party but cannot possibly cope with stewing "Apple Jack Frosted"—read on. The last recipes are for Howard Johnson's modern soda fountain desserts; yummy, elegant desserts that are easy to serve to any number.

APPLE JACK FROSTED

Core and quarter 6 tart green apples and stew in a syrup made of 2 cups water, 1 cup sugar, 1 teaspoon ground cloves, ½ teaspoon ground allspice and 1 lemon sliced thin. Simmer until tender. Let cool, taste for sugar, and add enough sugar to sweeten to taste, simmering 3 minutes more. Add ¼ teaspoon powdered cinnamon, ½ teaspoon ground ginger and 2 teaspoons vanilla extract. Slice 2 bananas into the apples and serve 2 tablespoons to the serving. Top with the same quantity vanilla ice cream.

CALCUTTA LEMONADE

Pare thinly the peel of 2 dozen fresh lemons, and let them stand in the juice of all the lemons for 12 hours. Add two pounds granulated sugar and stir to dissolve. Add 1 quart good dry sherry and three quarts milk. When cold, strain through a single layer of cheese cloth and serve with chopped ice and a straw. Serves 16.

CLARET CUP

Claret Wine 1 quart
Rhine Wine 1 pint
Sherry or Catawba Wine (if available) ½ pint
Sugar ¼ pound
Water 1 cup
Lemons 2 sliced thin

Let sugar, water and lemons stand for 1 hour at room temperature. Add wines. Stir well. Pour into a punch bowl filled with a block of ice. Add soda to taste—about 1 quart. Top each glass with a sprig of mint. Serves 10.

CLARET PUNCH

Tea, freshly brewed 1 quart
Sugar 1 cup
Brandy 1 pint
Saint Croix Rum ½ cup
Claret Wine 2 quarts
Crushed Pineapple 1 can
Orange 1 sliced thin
Lemon 1 sliced thin
Soda 1 quart or more to taste

The tea should be fairly strong: ¼ cup loose tea or 12 bags to 1 quart boiling water. When the tea has cooled to room temperature, add sugar and stir to dissolve. Combine tea, brandy, rum, wine and crushed pineapple. Chill thoroughly. When ready to serve, pour into a large punch bowl in which you have placed a cake of ice. Float sliced orange and lemon on top and add soda. 24 servings.

EGG LUNAR BLEND

Take 2 mixing glasses. Break an egg, putting the yolk in one glass, the white in the other; into the glass with the yolk, add 1 ounce cherry syrup and some cracked ice, shake, add small quantity soda and strain into a glass. In the other mixing glass add 1 ounce plain cream to the egg white and beat with spoon until well whipped. Add ½ ounce lemon syrup, then transfer it all into a shaker and add soda; float on top of the glass containing the yolk and cherry. Serve with two straws.

FRUIT PUNCH

Claret Wine 2 quarts
Crushed Pineapple 1 can (2 cups)
Lemons juice of 6
Oranges juice of 3
Sugar to taste
Champagne 1 quart
Soda 1 quart

Combine claret, pineapple and fruit juices. Sweeten to taste with sugar. Pour into punch bowl over a cake of ice. Add champagne and soda, and float slices of orange and lemon over the top if desired. Serves 10.

ICED TEA

Over 10 tablespoons of the best Ceylon tea, pour 1 gallon boiling water. Stand for five minutes and strain off the tea leaves. When cold, pour into a large punch bowl which contains a large piece of ice, add sugar to taste, squeeze in the juice of 6 lemons, and slice two lemons into the bowl. Serves 16.

REMEMBER . . .

"Paper napkins cost only one-seventh of a cent. Occasionally they are much wanted. Keep a few on hand."

MAPLE PARFAIT

If you follow these directions carefully, you will have the most perfect thing ever set out on a soda counter.

Eggs, yolks only 4
Maple Syrup ¾ cup
Heavy Cream, well whipped 1 pint

In the top of a double boiler combine beaten egg yolks with the maple syrup. Stir over the boiling water until mixture thickens and coats a spoon. Remove from fire and cool. Fold whipped cream into the cooled mixture and turn into a mold. Cover tightly and freeze without stirring for 4 or 5 hours. Serve in stem glasses with a spoonful of whipped cream. Serves 6.

ORANGEADE

Orange or Orangeade Syrup 1 pint
Soda Water 5 quarts
Orange 1 sliced thin

In a large punch bowl put a cake of ice. Add orange or orangeade syrup and soda to taste. Float orange slices on top and dip serving out with ladle. This makes a delicious drink; a big seller and much profit. Serves 16.

REGENT PUNCH
THE MOST DELICIOUS PUNCH MADE

Grate the rind of 2 lemons and 4 oranges. Add 2 cloves, 2 teaspoons of vanilla extract and a stick of cinnamon. Over all pour 1 pint hot simple syrup (2 parts sugar to 1 part water). Let stand for several hours and then strain into a punch bowl. Add the juice of 16 lemons (1 pint), 2 quarts rum and more syrup if desired. Add a block of ice and soda to taste—about 2 quarts. Serves 20.

HOWARD JOHNSON'S
MODERN PARTY DESSERTS

BRAZILIAN COUPE

Top a scoop of chocolate ice cream with a scoop of coffee ice cream, and over all pour 1 ounce crushed pineapple and pineapple syrup. Add a generous dollop of whipped cream and serve.

HELEN COUPE

Over a scoop of vanilla ice cream place one half peeled fresh pear. Over all pour 2 ounces hot chocolate-fudge sauce and garnish with sprinkles.

JANET COUPE

A scoop of pistachio ice cream tops a square of coconut cake. Add a spiral of whipped cream and garnish with walnuts or shredded coconut.

ISLAND COUPE

Top a scoop of coffee brandy ice cream with a spiral of whipped cream, dust with cinnamon powder and garnish with walnuts.

MARIANNE

On a scoop of vanilla ice cream place a square of sponge cake and top with blueberry sauce. Add a generous dollop of whipped cream and serve.

MELBA COUPE

In an ice cream dish place two squares of sponge cake and, between them, a scoop of vanilla ice cream. Slice fresh peeled peach over the ice cream and top with raspberry syrup.

"It is estimated that
in every 10 glasses of soda water sold
there are—

6 glasses Fruit Syrups with Ice Cream @ 2½¢	$.15
3 glasses Phosphate and Plain Soda @ 1½¢	.04½
1 glass Plain Soda Water @ ½¢	.00½
Making 10 glasses cost	.20
Average cost per glass	.02"

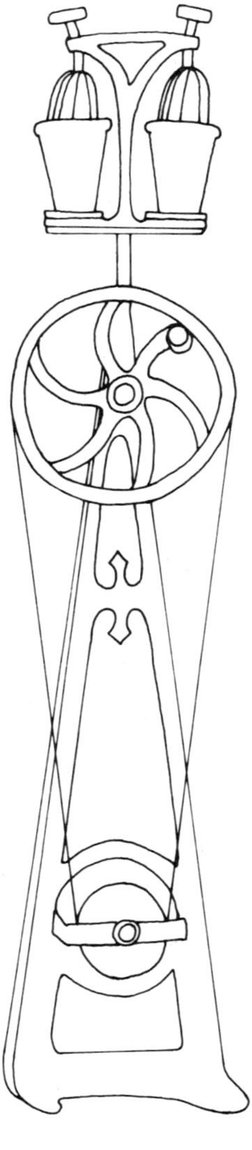

The lightning shaker

CONSUMER GUIDE

A guide to some of the syrups and sundries available for the concoction of old time soda fountain recipes. Most of the companies noted distribute their products nationally; in addition, supermarkets and specialty shops in your area may be stocked with the products of local importers and/or manufacturers.

COFFEE-TIME PRODUCTS: Manchester, New Hampshire
coffee syrup

FERRARA CONFECTIONARY COMPANY:
New York, New York
coffee syrup, lime syrup, orange syrup, orsata syrup (orgeat)

FLAVOR MILL COMPANY: Los Angeles, California
ginger extract, nectar extract, pistachio flavor, rose flavor, violet flavor

FOX'S-U-BET: Brooklyn, New York
cherry syrup, kola syrup, lemon-lime syrup, orange syrup, raspberry syrup, strawberry syrup, vanilla syrup

FRIMOST: Hamburg, West Germany; distributed by
A&A Food Product Corporation, New York
blueberry syrup, raspberry syrup, red currant syrup, strawberry syrup

GIFFARD: France
orangeade syrup, mint syrup

HAIN PURE FOODS: Los Angeles, California
apricot syrup, grape syrup

HARRY and DAVID: Bear Creek Orchards, Medford, Oregon
4 Fruits Vinegar (apple, grape, pear & pineapple), pear vinegar

HERSHEY CHOCOLATE: Hershey, Pennsylvania
chocolate syrup

KNOTT'S BERRY FARM: Buena Vista, California
apricot syrup, blackberry syrup, strawberry syrup

KRAFT FOODS: New York, New York
chocolate syrup, chocolate fudge syrup, pecan topping, pineapple topping, strawberry topping, walnut topping

McCORMICK SPICE COMPANY: Newark, New Jersey
banana extract, pineapple extract, root beer extract, sour salt (citric acid), strawberry extract, walnut extract

S. S. PIERCE CO.: Boston, Massachusetts
claret sauce (syrup), fruit juices

SHERMAN FOODS: New York, New York
blackberry syrup, grape syrup, red raspberry syrup

SHMUCKER'S: Orrville, Ohio
apricot syrup, blackberry syrup, blueberry syrup, cherry topping, chocolate fudge topping, pineapple topping, strawberry topping, Swiss milk chocolate fudge topping

SPICE CLUB FOODS, INC.: Long Island City, New York
crême de apricot, crême de cacao, crême de menthe, crême de peach (non-alcoholic toppings and syrups)

VIRGINIA DARE: Brooklyn, New York
celery extract, coffee extract, peach extract, raspberry extract, a wide variety of syrups

JOHN WAGNER & SONS, INC.: Ivyland, Pennsylvania
claret sauce (syrup), sour salt (citric acid)

INDEX

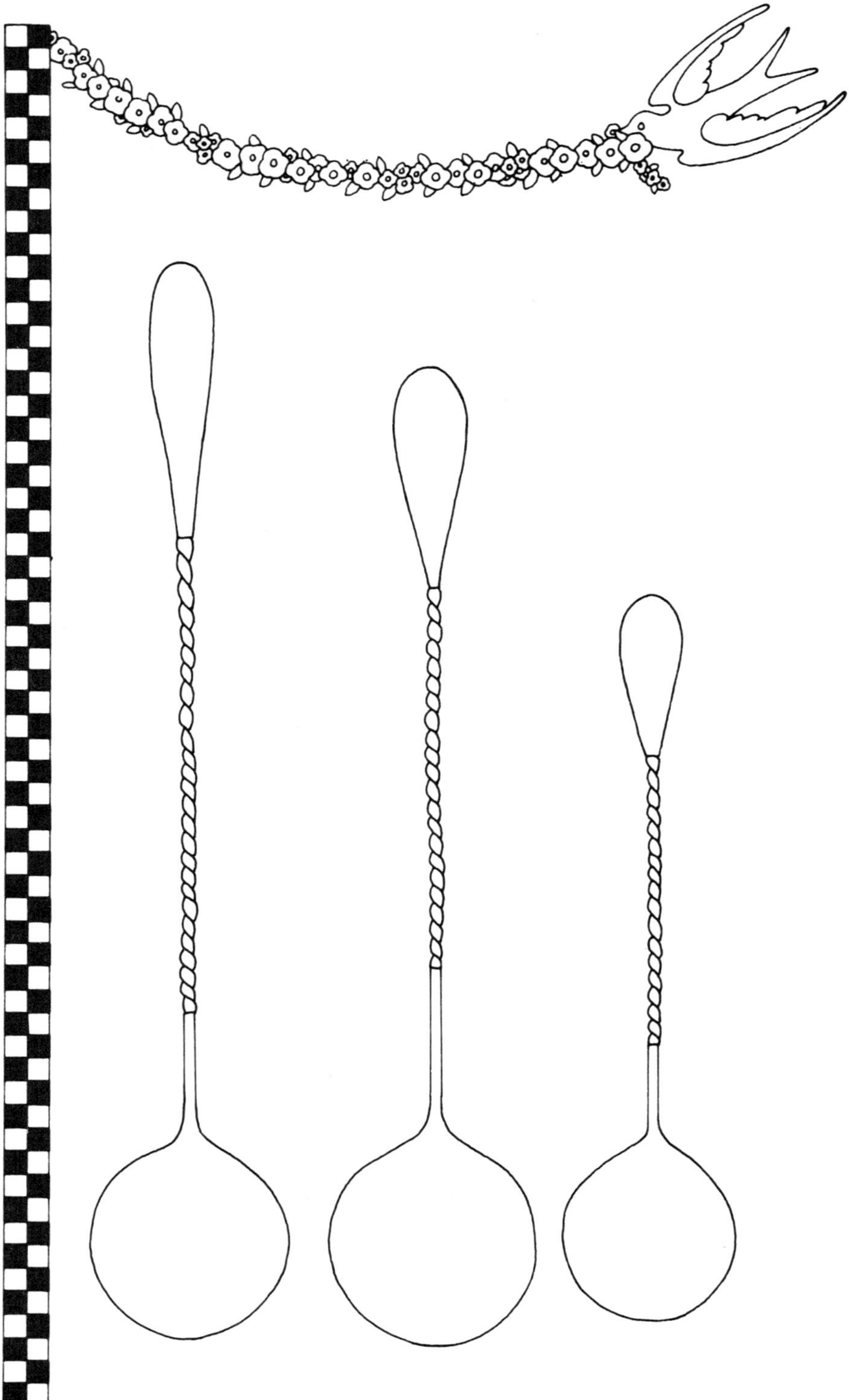

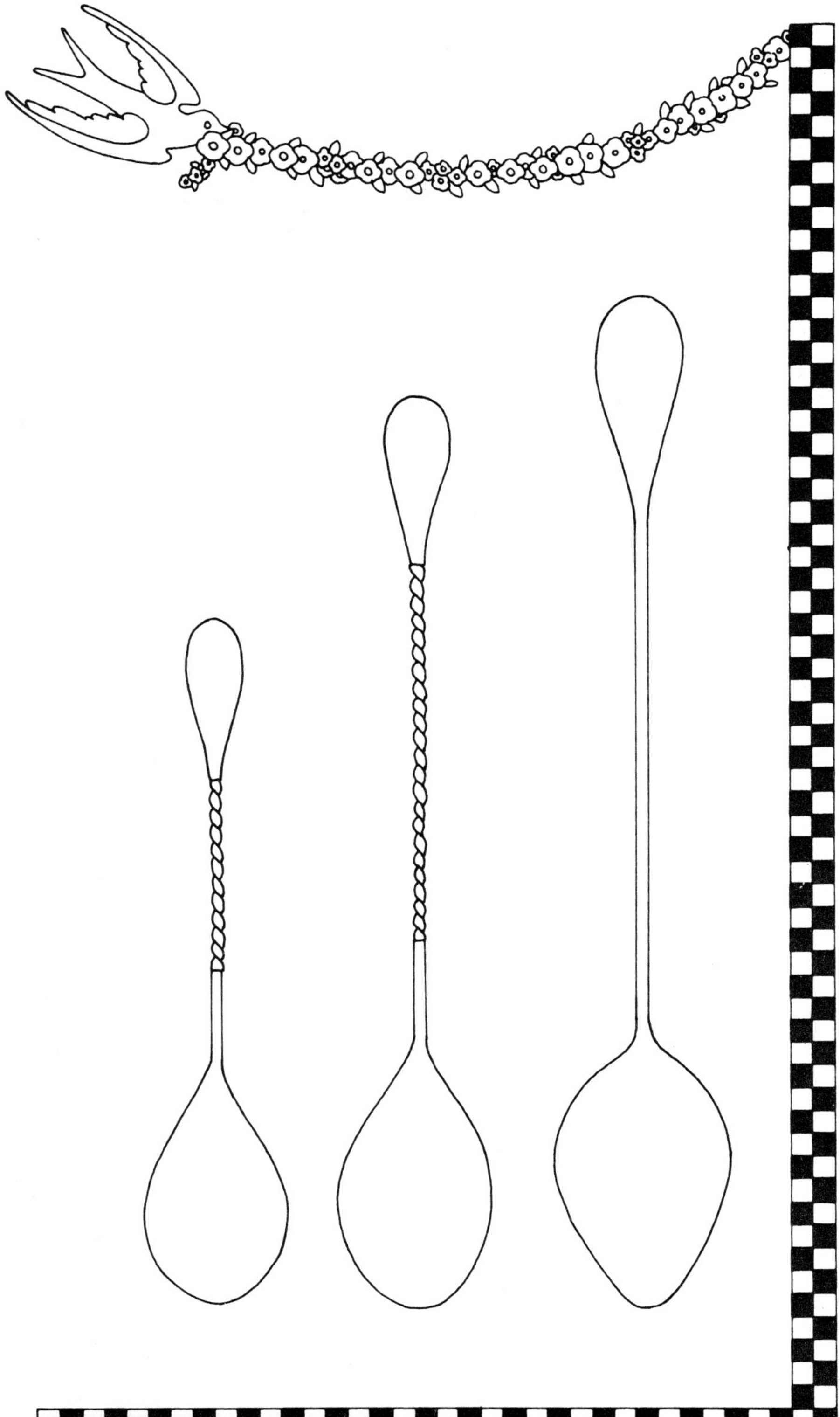

NOTES

"Don't use
too much syrup. It makes
a sickening drink."

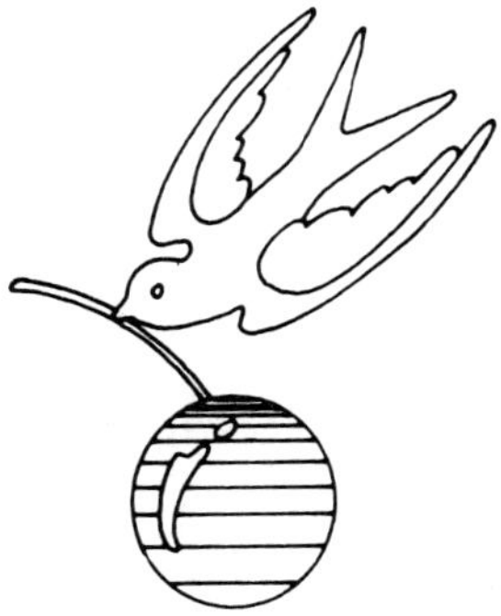

NOTES

"Don't turn your back
on a customer. Draw soda water
standing sideways."

NOTES

"The finest beverage
loses its charm if indifferently
served."

NOTES

"Use odd moments
to keep the
silver bright."

NOTES

"Be lavish in every direction—except as to quantity."

NOTES

"Never refuse anyone a
glass of water. A water customer today,
a soda customer tomorrow."

NOTES

"Don't handle
the rims of glasses with
your fingers."

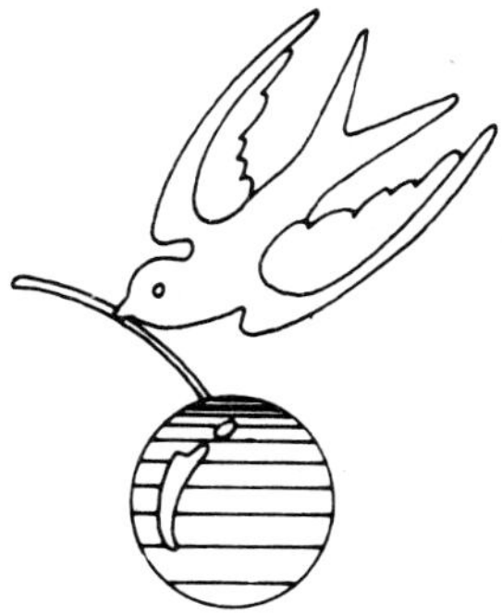

NOTES

"A final word:
an electric fan will keep away flies.
Think it over!"

The text of this book was set
in Bookman, and the text and cover were
printed and bound by the George Banta Company.
The cover and page borders were designed
by Barry Zaid of the Push Pin Studio,
and the illustrations are by
William Hunter.